Jo

SOLID MODELS

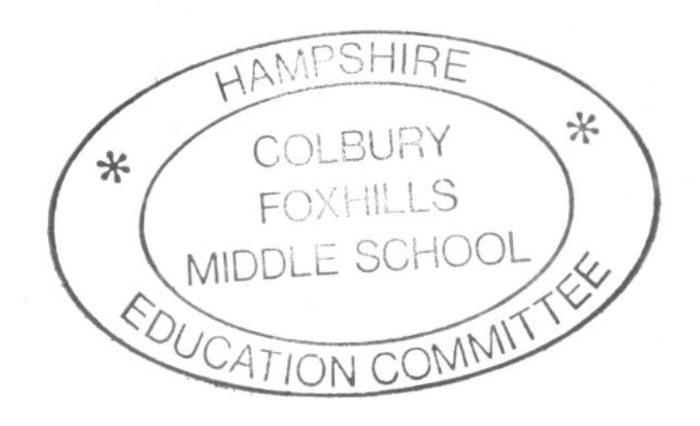

CAMBRIDGE UNIVERSITY PRESS

Published by the Syndics of the Cambridge University Press
Bentley House, 200 Euston Road, London NW1 2DB
American Branch: 32 East 57th Street, New York, N.Y. 10022

Library of Congress Catalogue Card Number: 66-16668

ISBN: 0 521 05745 0

First published 1967
Reprinted 1973

First printed in Great Britain by Jarrold & Sons Ltd, Norwich
Reprinted in Malta by St Paul's Press Ltd

The Regular Solids

Fig. 1

The solid in this photograph has the imposing name of SMALL STELLATED DODECAHEDRON but is not nearly so complicated as it sounds. In fact two people would be able to make it in just a few lessons.

You will need at least one large sheet of card, some quick-drying adhesive, some plain drawing-paper and geometry instruments. For cutting collect together scissors, a sharp knife and, if possible, a drawing-board and a metal-edged rule.

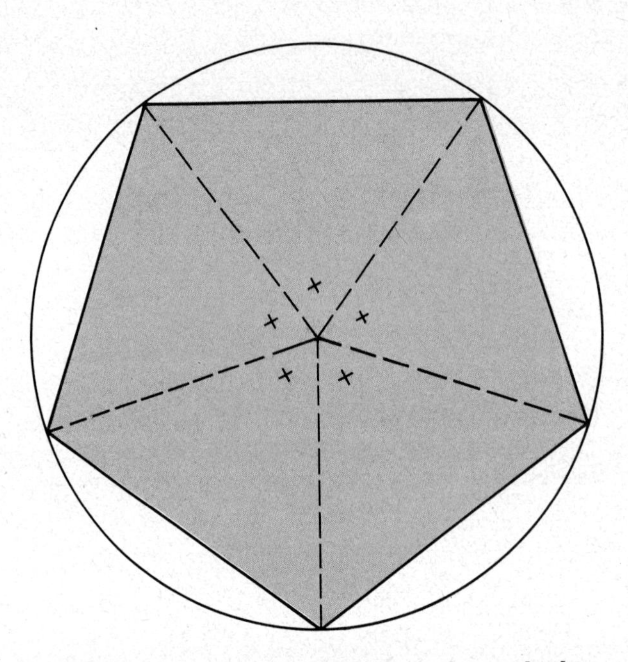

Fig. 2. A regular pentagon in a circle

Fig. 2 shows a five-sided figure, a pentagon. It is called a REGULAR pentagon because all the sides are the same length and all the angles are the same size.

All the angles at the centre (they are marked with a cross) are the same size and together they make 360° (one complete turn). Draw a circle about three inches across and work out what each of the five angles should be. Then use a protractor to find the positions of the five dotted radii. Join up their ends to form the pentagon. Check that all the sides are the same length.

If you were to extend all sides until they met another you would get a pentagonal star, called a PENTAGRAM. Fig. 3 shows this but it has been reduced in size when compared with the previous diagram.

Extend the pentagon you have drawn to make a star and keep it carefully as you will need it later on as a pattern or template for the shapes you will be drawing on card.

Fig. 3. A regular pentagram (five-pointed star)

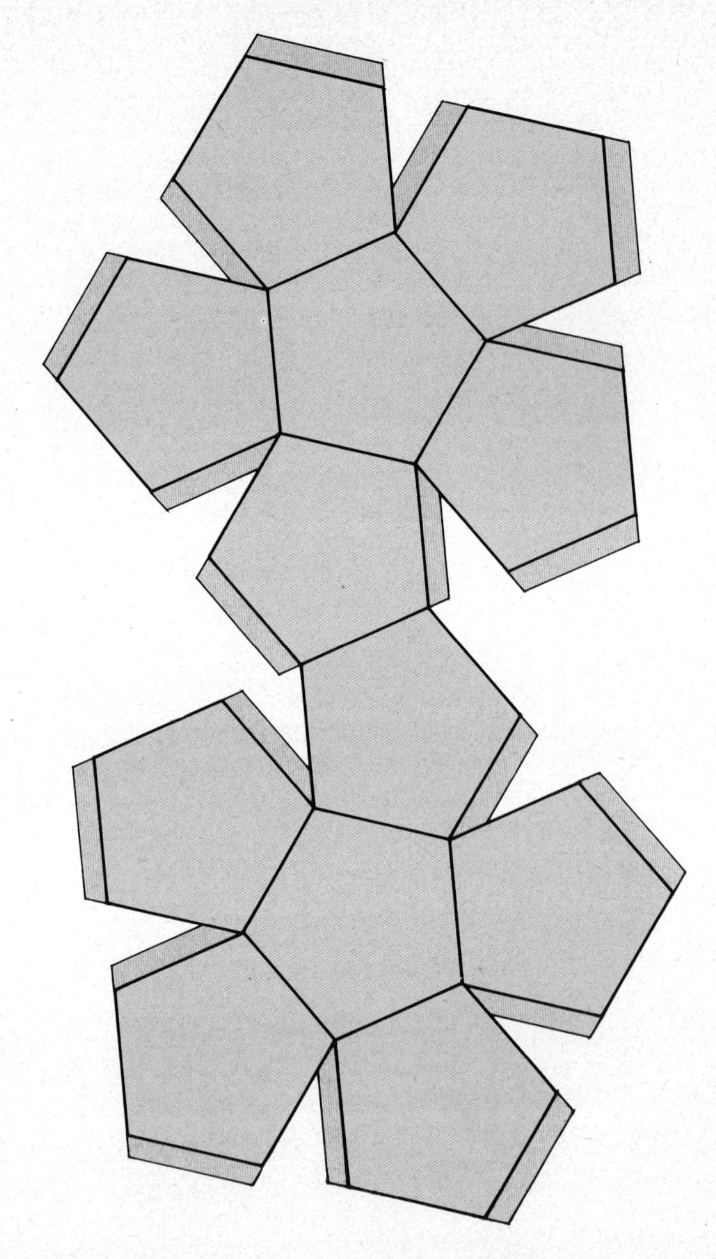

Fig. 4. Plan made of twelve regular pentagons ready to form a dodecahedron. The tabs for sticking are shaded

Fig. 4 shows you how to arrange twelve pentagons in the plan of a DODECAHEDRON.

A dodecahedron is a solid with twelve surfaces (called FACES). It will form the core of your model. The tabs on the edges are shaded in Fig. 4 and are used to stick the faces together. Take a tracing of the pentagon you drew and use it to draw the plan on thin card, then add the tabs.

To make the plan into a solid shape you will have to bend the card. You will be able to do this neatly if you run a sharp knife against a metal rule on the lines of the plan and then bend the surfaces *away* from the cut. This process is called scoring. Be careful that you don't cut right through the card. Now cut out and score the plan and use a quick-drying adhesive or cement to stick the tabs inside.

Fig. 5. A completed dodecahedron

The photograph in Fig. 5 shows the complete dodecahedron, but eventually this solid will be hidden because a five-sided pyramid will be stuck on each face (Fig. 6).

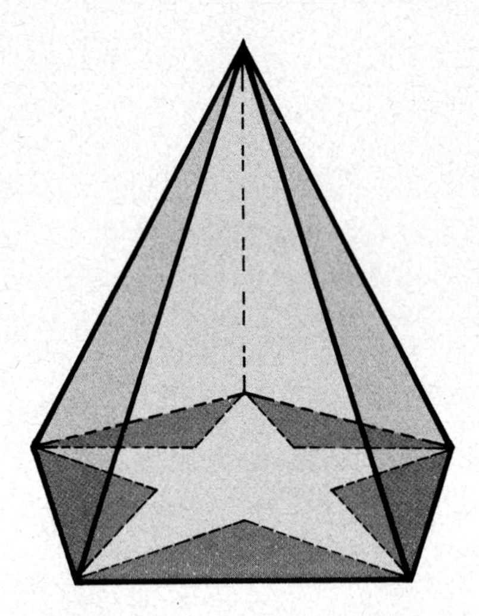

Fig. 6. Perspective drawing of a pentagonal pyramid

The plan for these pyramids is in Fig. 7. You see that they are made of five triangles side by side, with some tabs of course. These triangles are the same size as those on the pentagonal star (pentagram) which you drew earlier, so you will need to go back to that for a tracing of a single triangle. By repeating this triangle you will be able to draw the plan. Make enough copies of this on thin card. Cut and score them. Stick the long tab to make a pyramid and carefully bend the bottom tabs under and then stick them to the faces of the dodecahedron.

Your model is now complete unless you would like to paint it, but before you do this you should study the solid you have made in some detail.

Can you find any pentagrams like the one you drew at first (Fig. 3)? If you count them you will see that there are twelve but the central pentagon of each one is hidden by a pyramid. The twelve hidden pentagons form, as you already know, a dodecahedron. This accounts for part of the name of the

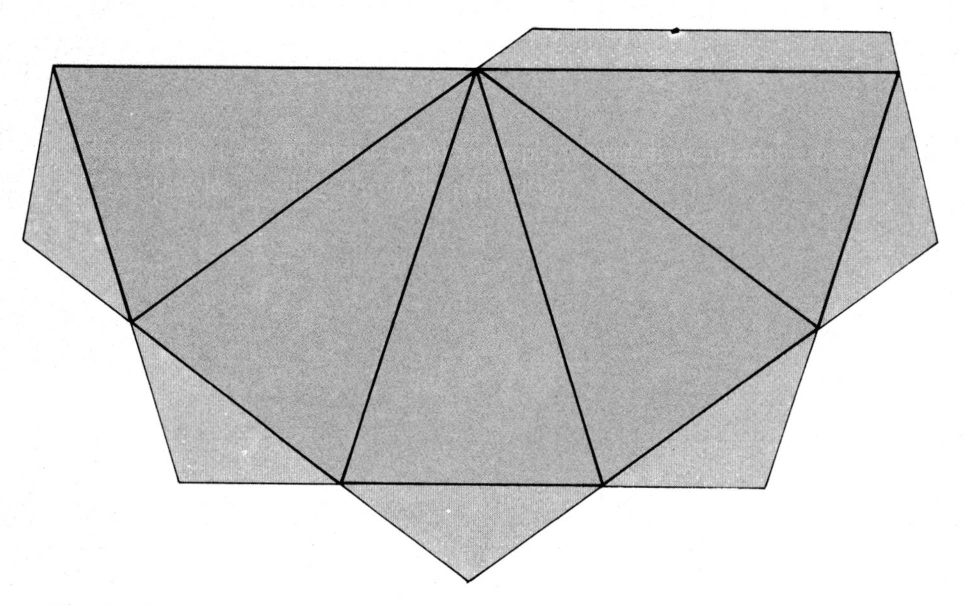

Fig. 7. Plan of pentagonal pyramid

shape. The word 'stellated' means 'starred'. So the entire name means starred and twelve-faced. The twelve faces of the solid are the pentagrams and not the hidden pentagons; they interlock in quite a complicated way which you could show up by painting each pentagram a different colour. However, it is possible, using only four different colours, to paint the solid so that no two faces of the same colour meet. If you plan to do it this way, instead of using twelve different colours, you should discuss it with your teacher who will help you to decide where to use each colour.

The solid you have just made is one of the regular solids. All its faces are the same shape and all the star points (each is called a VERTEX) are identical.

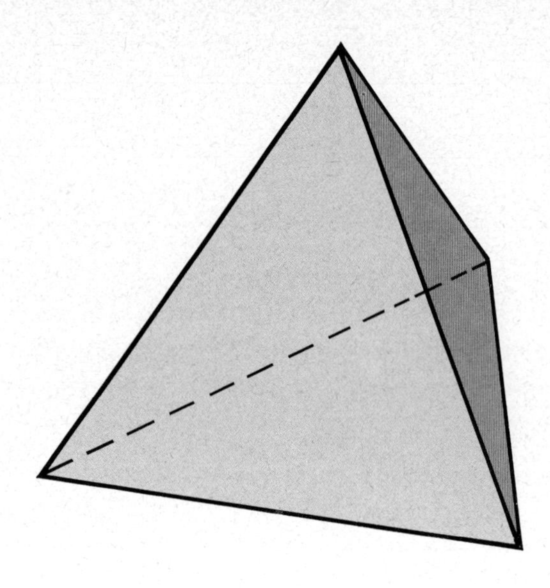

Fig. 8. Tetrahedron

One of the simplest regular solids is the TETRAHEDRON. Fig. 8 shows a sketch of one as though it were transparent. Make a skeleton model of this using straws and modelling clay.

How many faces has this shape got? They are all called equilateral triangles, meaning that they are triangles with all sides the same length. Find out how to draw one exactly if you do not already know.

The corners of a solid are called its vertices. Write down the number of vertices of a tetrahedron.

Where two of the faces meet, a straight line, called an edge, is formed. Write down the number of edges of this solid. Use a good dictionary to find out which Greek words the name tetra-hedron comes from. Why does it have this name?

You now know how many faces there are and that they are all equilateral triangles. Decide what the plan for this solid should be like and draw it on card. The special name for such a plan is a NET. You have already seen the net of a dodecahedron in Fig. 4.

Now put tabs on the net, cut it out, score the edges and make it up into a tetrahedron.

Two other solids which you will easily be able to make yourself are the CUBE and the OCTAHEDRON.

Here is a sketch of a cube. What shape are the faces? How many are there? Draw a suitable net for a cube and make the solid from thin card.

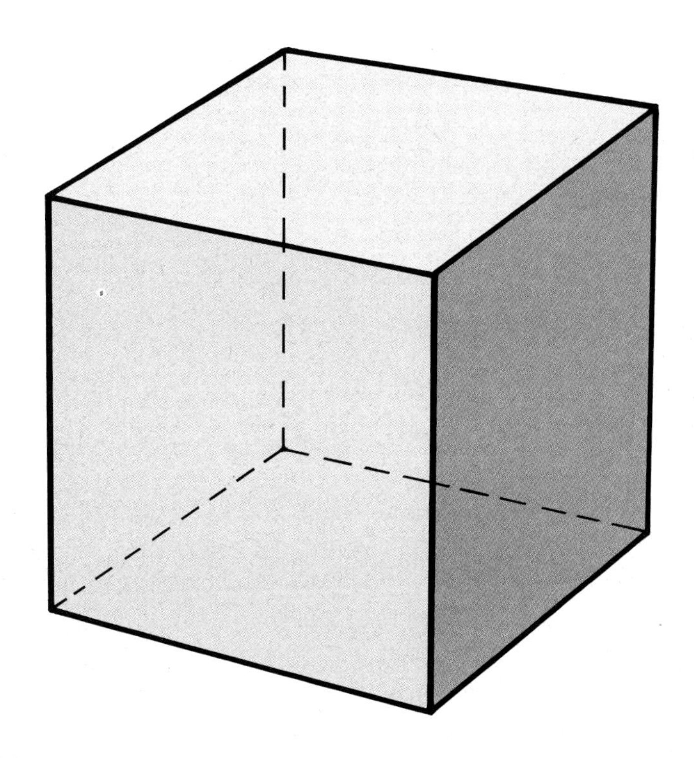

Fig. 9. Cube

The next diagram shows an octahedron. Make one from straws and Plasticine.

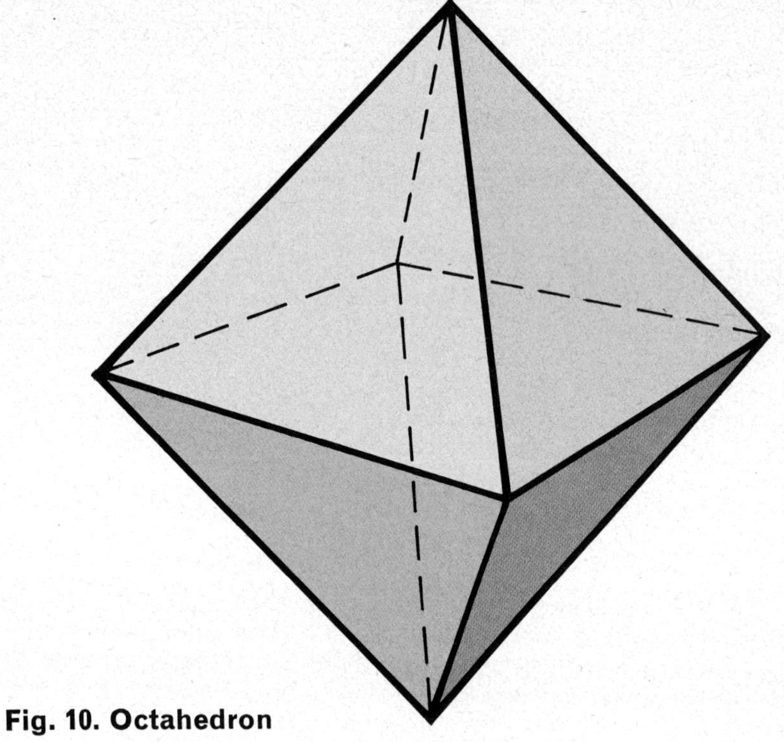

Fig. 10. Octahedron

Find out as many words as you can which begin with OCT. (Use your dictionary.) Write a list of them with their meanings. Explain how an octahedron gets its name.

As you see, all the faces are equilateral triangles. You could think of it as two four-sided pyramids together. You already know how to make a five-sided pyramid. Use this knowledge to help you design a suitable net for the octahedron and make it up into a solid.

There are several solids whose faces are equilateral triangles only. You have already met two of them. They are all known as DELTAHEDRA, because their faces are the shape of the Greek letter delta which is written △. A small delta is written δ so you see how we got our letters D and d.

There are five very simple regular solids and you have already made four, the tetrahedron, octahedron, cube and dodecahedron. Two of these are deltahedra on account of the shape of their faces. The only one of the five simplest regular solids you have not yet made is also a deltahedron. It has 20 equilateral triangles for faces and is called an ICOSAHEDRON. The net and tabs for this solid are drawn in Fig. 11. Make a large and accurate copy of this and construct an icosahedron.

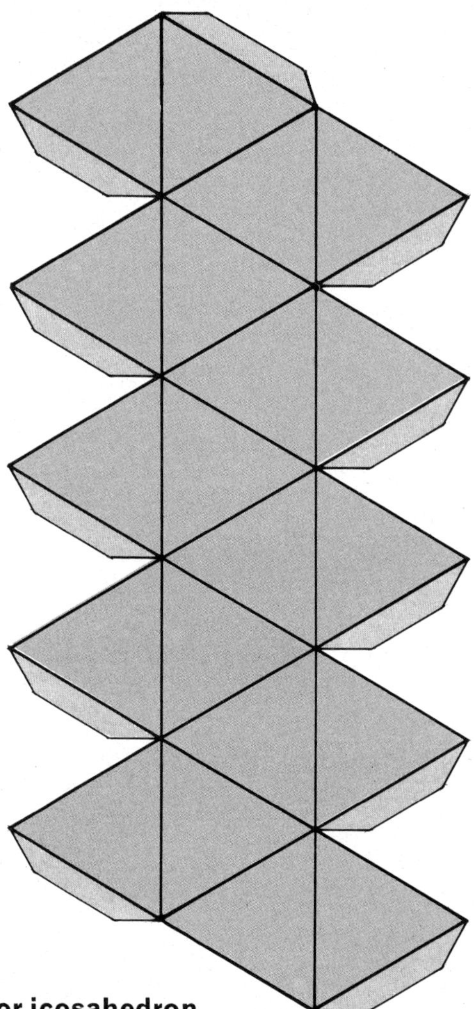

Fig. 11. Plan for icosahedron made of 20 equilateral triangles

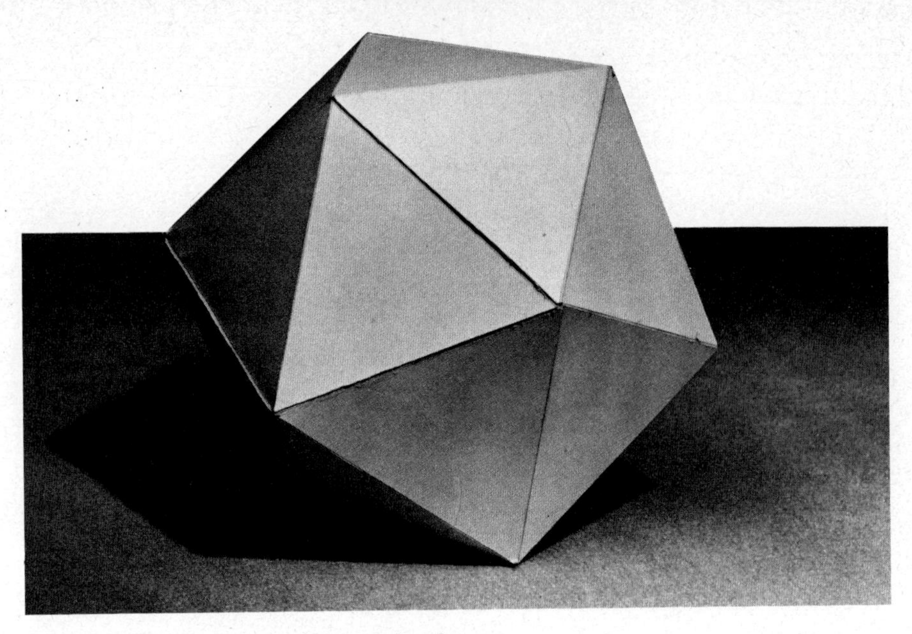

Fig. 12. A completed icosahedron

If you want to have a complete set of the five solids you will need to make another dodecahedron as your first was used inside another model.

Now collect together the five simple solids and after examining them copy out and complete this table.

NAME OF SOLID	NUMBER OF FACES	NUMBER OF VERTICES	NUMBER OF EDGES
Tetrahedron			
Cube			
Octahedron			
Dodecahedron			
Icosahedron			

If you study the table you will find some connection between the number of faces and vertices and the number of edges of each individual solid. Write down what you notice and ask your teacher if you are right. This relationship was studied by a mathematician called Euler (pronounced 'Oiler') who lived from 1707 to 1783. Find out something about his life and work by looking at a mathematical reference book or an encyclopaedia.

You have made only a few of the solids which you could make in this way. Some of the others are much more difficult to make but are very interesting and attractive to look at especially if well painted. The stellated dodecahedron which you made at first is another regular solid and has two close relations. You can find out much more about them if you look at a book called *Mathematical Models* by H. M. Cundy and A. P. Rollett. You will probably find that your maths. teacher has one or that there is a copy in the library.

Finally arrange your models in an attractive display. With each one put an information label giving its name, number of faces, edges and vertices, its net perhaps, and of course your own name and the date when you completed it.

Duality – Two by Two

Look again at the table on the opposite page. Do you notice anything about the numbers connected with the cube and the octahedron?

They have the same number of edges but their faces and vertices are interchanged. On account of this each one is known as the DUAL of the other. Fig. 13 illustrates this. You will see that the centre of each face of the cube has been made into a vertex of the octahedron. This can only be done because the numbers are the same. Fig. 14 shows it the other way round. This time the centres of the eight faces of the octahedron have become the eight vertices of the cube. If you had enough patience you could make a model showing one inside the other and another inside that until they were too small to make.

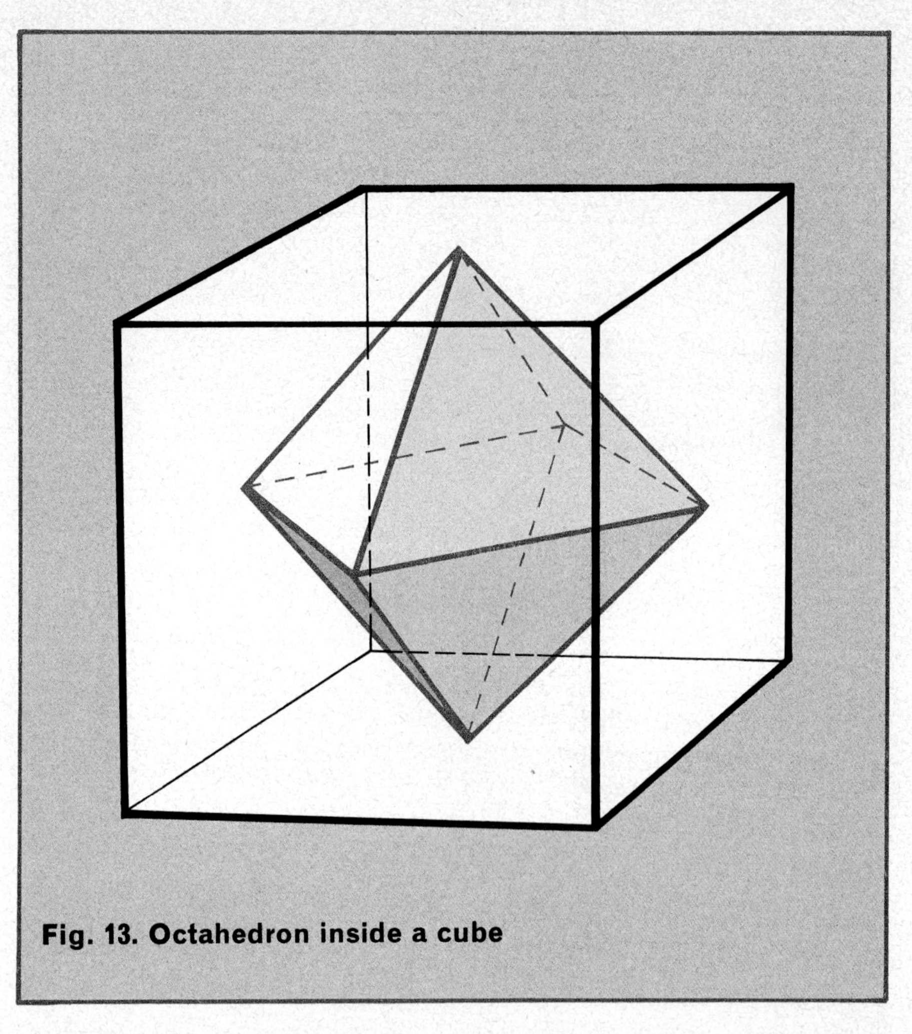
Fig. 13. Octahedron inside a cube

You can make the models of these dual solids very effectively by making the framework of the outer shape in $\frac{1}{8}$-inch balsa wood stuck with balsa cement. The frame is then covered with a skin of polythene, cellophane or plastic sheeting. You must remember to leave a gap to put the inner solid through. To find the measurements for the inner shape you should mark the centres of the faces as accurately as you can and find the distance from one centre mark to the other using callipers or dividers. Your teacher will show you how to do this. You can then make the dual solid from brightly painted

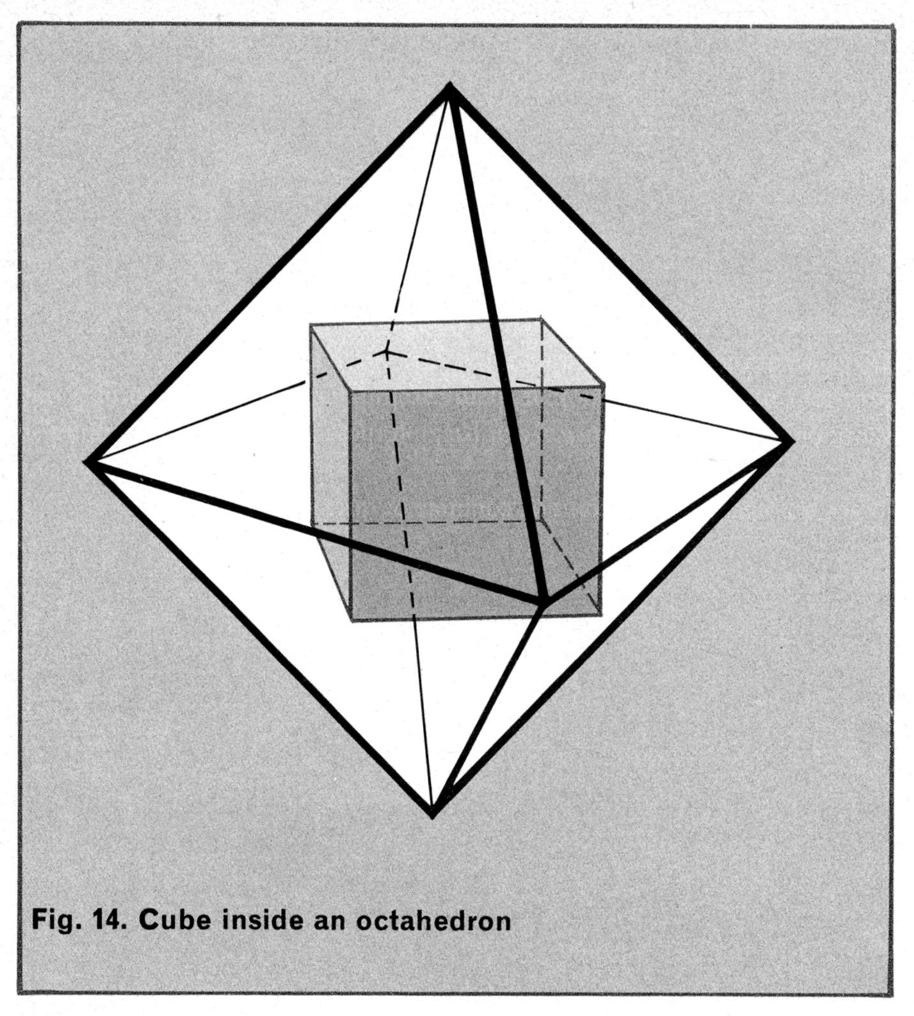

Fig. 14. Cube inside an octahedron

card, place it inside, stick the vertices to the centres of the transparent faces and then cover the last face of the outer shape.

Look once more at the table on page 14, to see which solid is the dual of the tetrahedron. You will notice that there are the same number of faces as vertices. If you interchange the numbers to find the dual it will make no difference, you still have a tetrahedron. For this reason we call it SELF-DUAL. Fig. 15 shows a sketch of this. Make a model as you did those in the last paragraph.

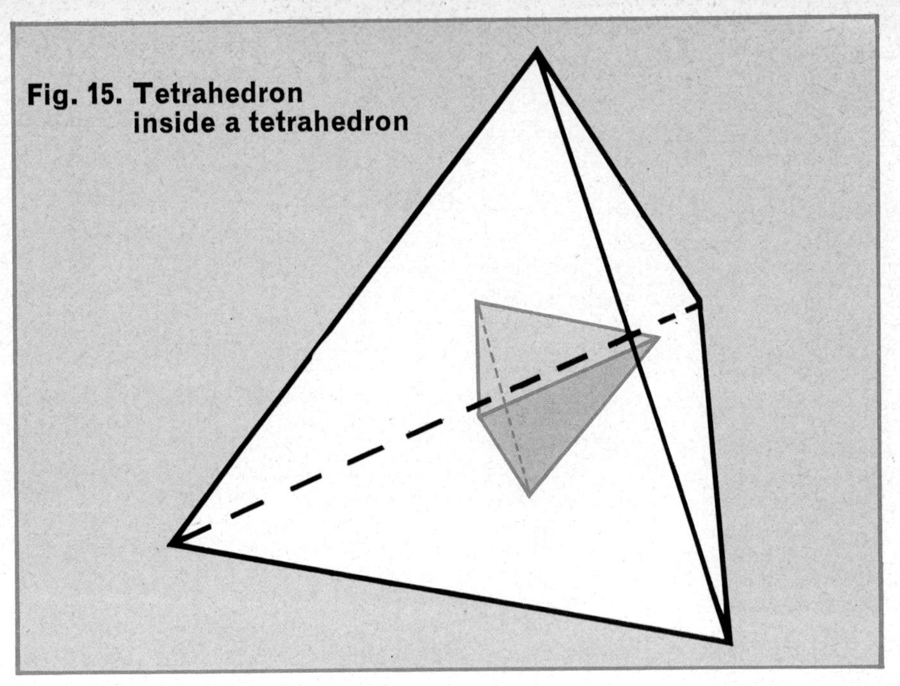
Fig. 15. Tetrahedron inside a tetrahedron

Lastly, look at the icosahedron and the dodecahedron and you will see that they are dual also. It would be difficult for you to make these as you have made the others but you can demonstrate it quite easily if you go back to your first model, the small stellated dodecahedron. You remember that the core of this is a dodecahedron with twelve faces. Each face has a star point. Decide on a way to join the star points; perhaps by thin balsa wood strips or straws or bright string or wool sewn through the tips. Look at the skeleton shape which you have made in this way. Write an explanation saying why this has turned out to be an icosahedron. The cover picture shows the model you have just made.

If you were to make a stellated dodecahedron from thin balsa wood and to join up the vertices with strips of wood then you could show up the inner core and the outer shape using bright poster paint.

There is a book with some really good photographs of dual solid models which you might find in the library or be able to borrow from a teacher. It is called *Mathematical Snapshots* by B. Steinhaus. You will probably be interested in many of the other pictures and ideas in this book.

Mathematical Skeletons

Some of the most attractive models are made by using thin strips of balsa wood held together with balsa wood cement and painted really brightly with poster paints. They are so light that they can be hung up with cotton and they move slowly round in the air. If you can suspend them under an electric light so much the better as the light makes the paint show up particularly well.

Here are a few examples which you will be able to make. Once you have got the idea you will be able to invent your own.

The model in Fig. 16 is made up of strips in two sizes, one size is twice the length of the other.

When you have finished the model answer these questions. What is the outer shape? What is the inner shape?

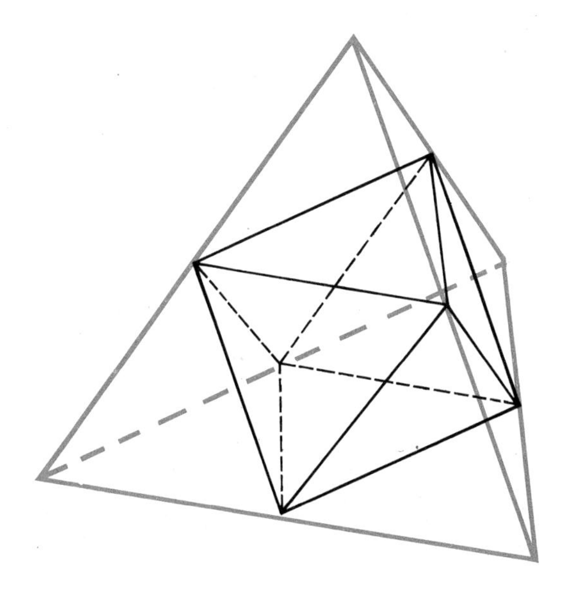

Fig. 16

By referring to the table you filled in about the regular solids write a short explanation of why the inner shape has turned out as it has.

Now make a second tetrahedron round the inner shape. Try it out with straws and Plasticine first if you like.

When you have done this you will have made a new outer solid which is made up of two interlocking tetrahedra. It is called STELLA OCTANGULA. Why? If you have forgotten what 'stella' and 'oct' mean you will have to look back in the book and your notebook, then you will realize how this solid gets its name.

You can also make a stella octangula by sticking a tetrahedron on to each face of an octahedron. Make one in this way and paint it to show the two interlocking shapes (Fig. 17).

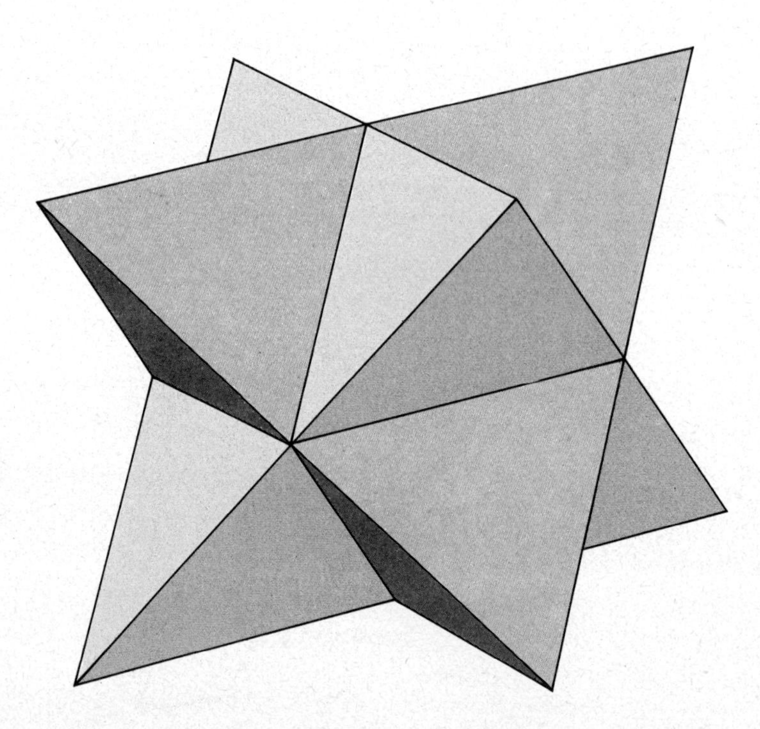

Fig. 17. Stella octangula shaded to show two tetrahedra

The next two skeletons are based on a cube so have two skeleton cubes ready.

For the first imagine the vertices of the cube numbered as Fig. 18.

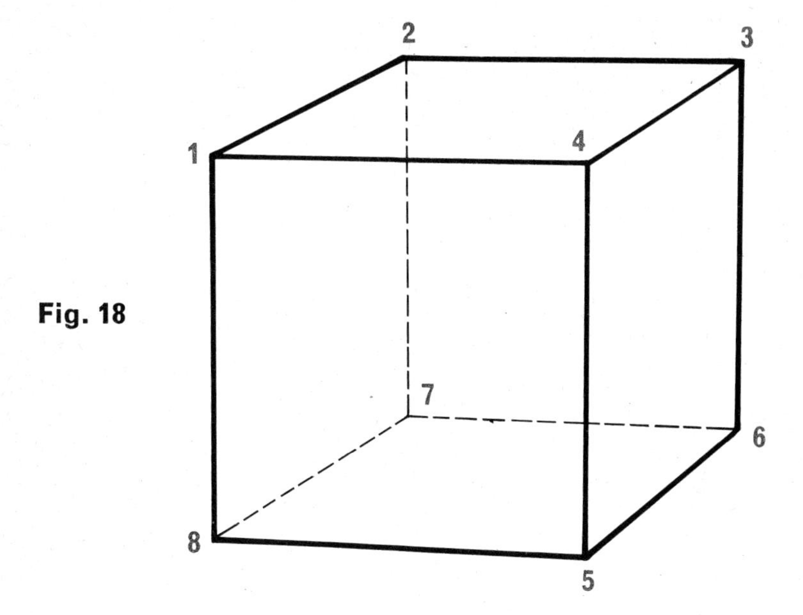

Fig. 18

Now join all the odd numbered vertices together in some way. How many have you joined? How many new lines have you made? What shape has been formed?

Now join the even vertices—you should have the same shape again. These two shapes together form a stella octangula once more but you got it in quite a different way from before. Compare it with the previous models.

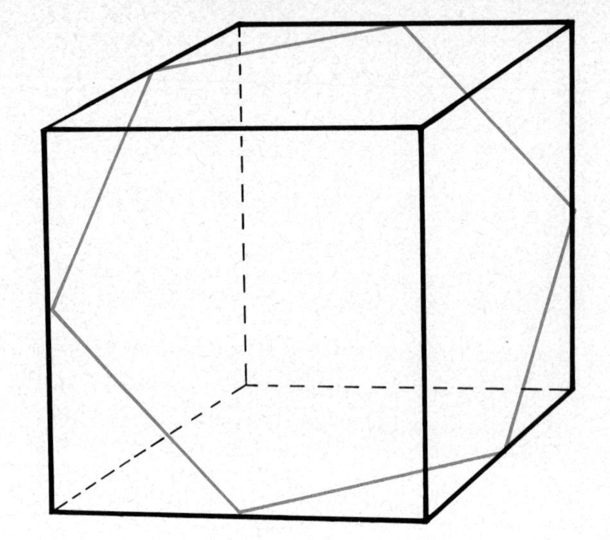

Fig. 19

The second model based on the cube begins in Fig. 19. You will see that the midpoints of certain edges have been joined.

Find out the name of the flat shape which has been made inside. How many others the same as this could be put into the cube? You could illustrate your answer with sketches.

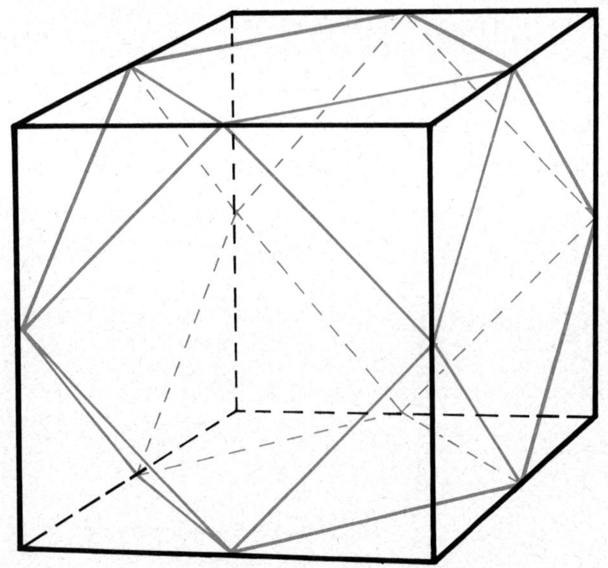

Fig. 20

Fig. 20 shows the shape you would get if you were to put in all the possible hexagons. It can also be made by taking the octahedron as the basic shape (Fig. 21).

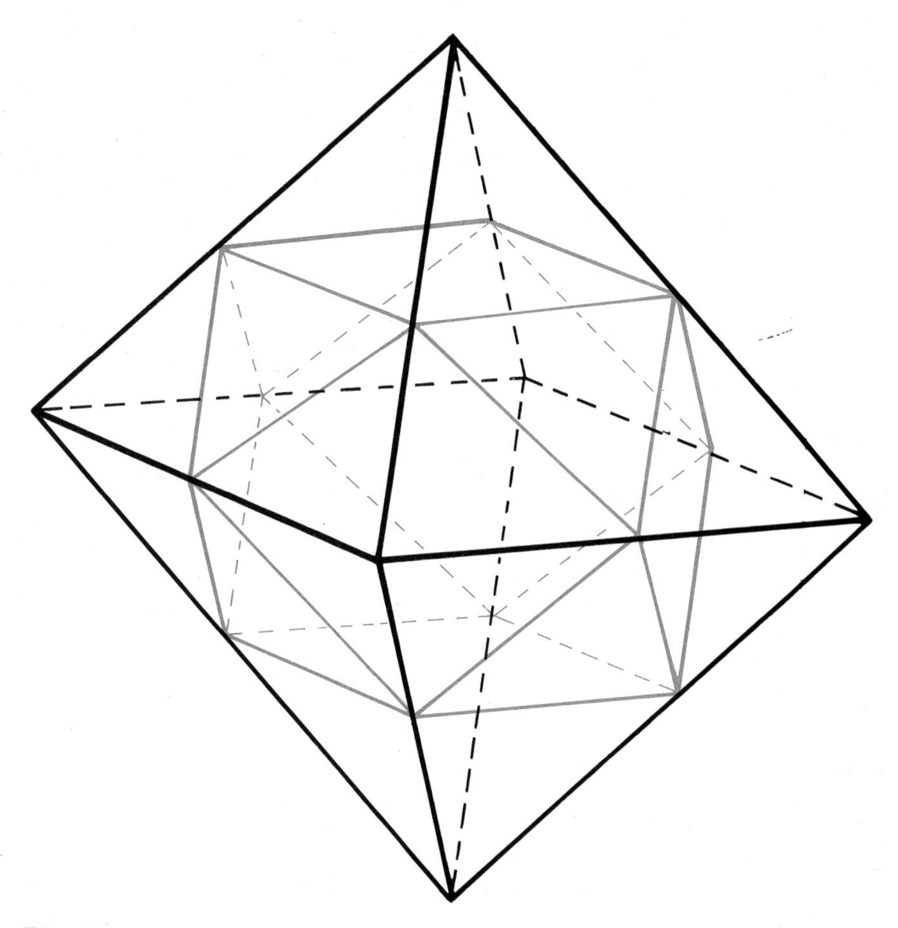

Fig. 21

On account of its connection with the cube and the octahedron this solid is known as a CUBOCTAHEDRON.

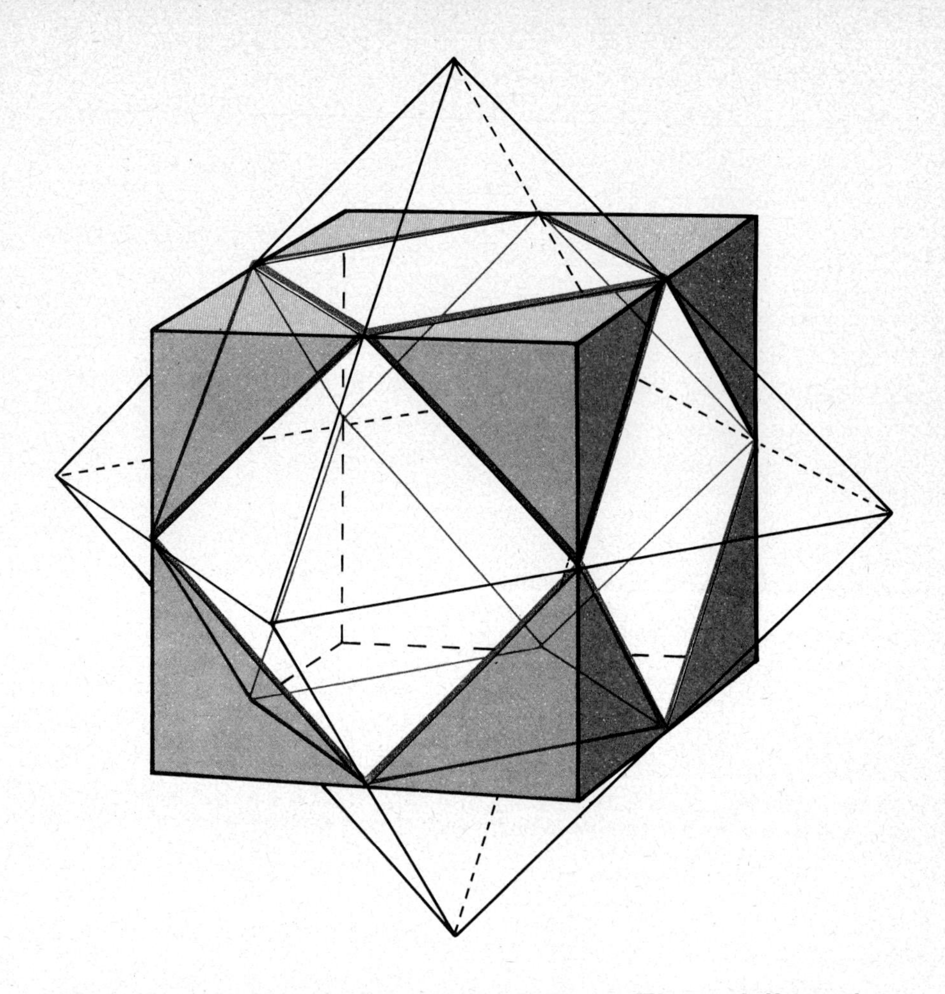

Fig. 22. Cube interlocked with octahedron. The red lines show where the two solids meet and form a cuboctahedron

It is the shape which is common to the cube and the octahedron and we say it is the INTERSECTION of these two shapes. Fig. 22 illustrates this idea, picking out the shape made where the cube and the octahedron meet.

When you have finished the models of Fig. 20 and Fig. 21 look carefully at the inner shape, the cuboctahedron, and answer these questions.

How many faces has the cuboctahedron?

They are of two types. What are these types?

How many of each kind are there?

How do these numbers and shapes compare with those of the cube and the octahedron themselves?

Now draw a net for making the cuboctahedron. When you have made up the solid, paint the faces using two colours.

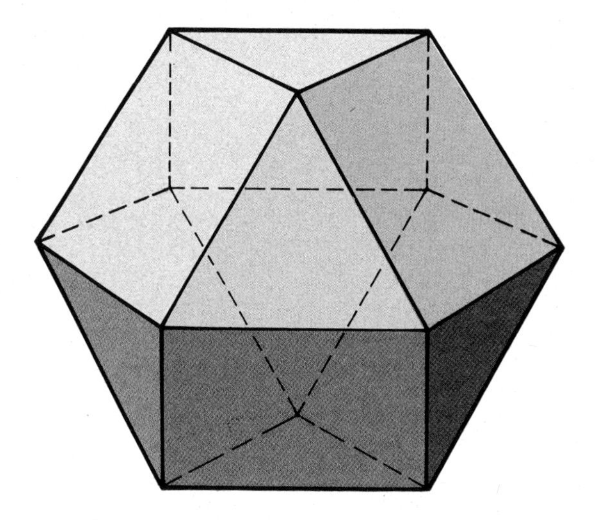

Fig. 23. Sketch of cuboctahedron

Here is a final problem about cuboctahedrons for you. Make a set of pyramids which when placed on the triangular faces make it into a complete cube. Then make another set which when placed on the square faces make it a complete octahedron.

Semi-regular Solids

You will remember that the first shapes you made were called regular solids because all the faces were the same and each vertex had the same number and arrangement of faces meeting there.

The cuboctahedron (page 25) has two different types of face and so cannot be regular. You will notice, however, that all the vertices are made up in the same way with two squares and two equilateral triangles meeting at each one. For this reason the solid is called SEMI-REGULAR. Remembering that complete regularity means having identical faces *and* identical vertices you can understand why it is only semi-regular.

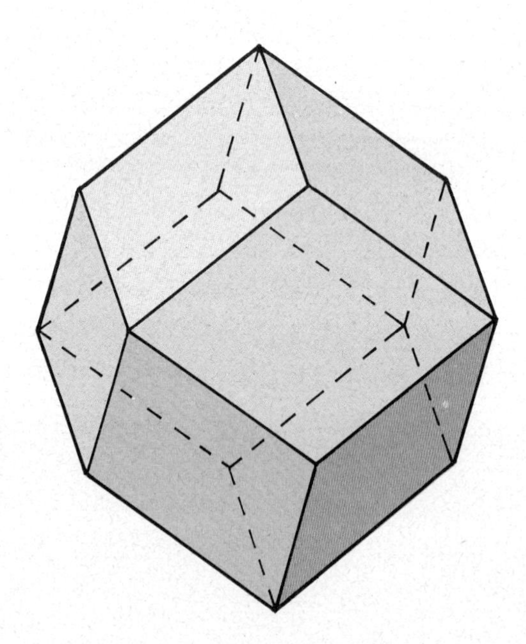

Fig. 24. Sketch of rhombic dodecahedron

The solid in Fig. 24 is called a RHOMBIC DODECAHEDRON. Rhombus is the mathematical name for what you recognize as a diamond shape. Does dodecahedron tell you how many faces to expect?

This model is also semi-regular. This time all the faces are the same and it is the vertices which are different. When you have made one you will see how this is so. All the faces are made from the same rhombus and you must draw it so that the smaller angle is as near to $70\frac{1}{2}°$ and the larger angle to $109\frac{1}{2}°$ as possible (Fig 25).

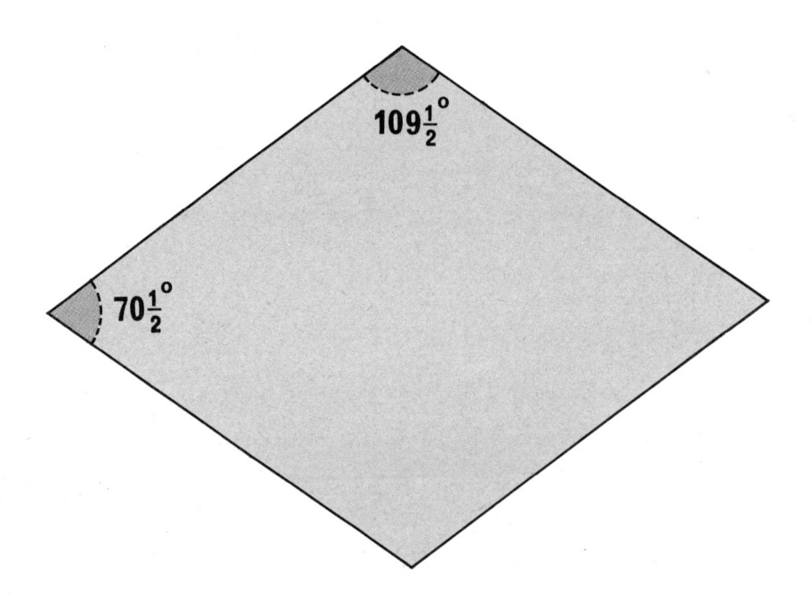

Fig. 25. The rhombus which forms the faces of this solid

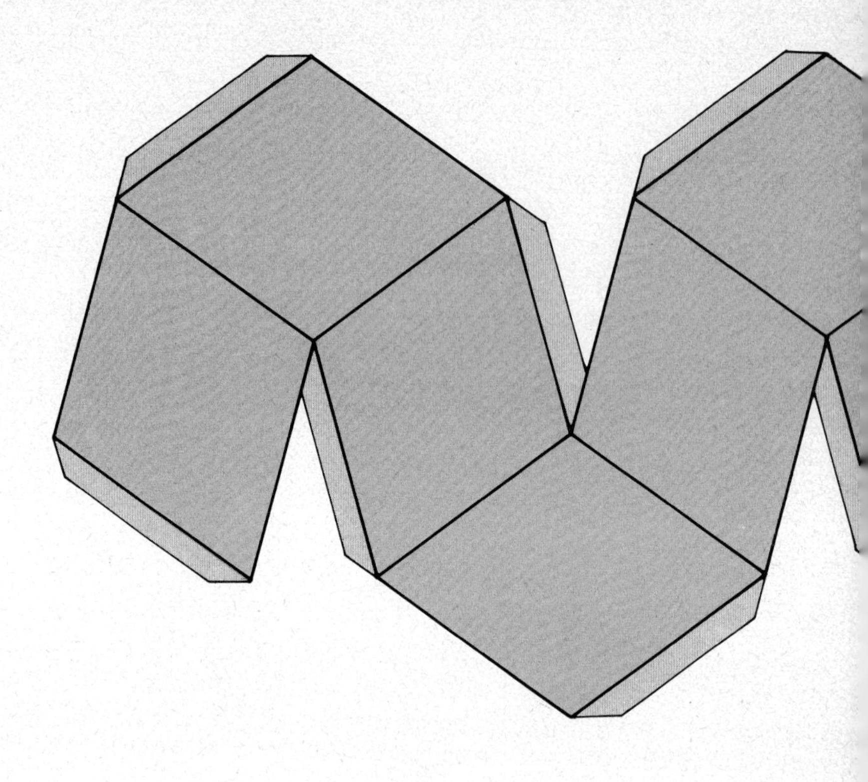

Fig. 26. Net of rhombic dodecahedron

Fig. 26 shows you one way of arranging the shapes to form the net.

When you have completed the model look at the vertices carefully. They fall into two types. Describe each type and mark each with a different colour.

Now with a ruler and a coloured pencil join up all the vertices which are one colour by drawing lines across the faces.

How many lines have you drawn?

How many vertices have you joined?

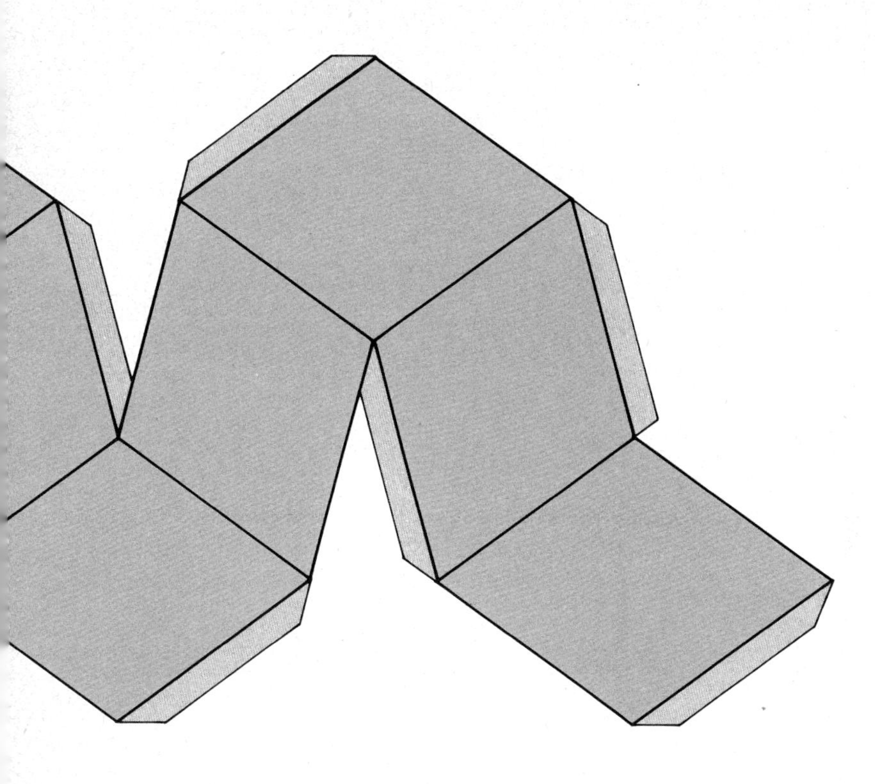

What shape has this number of vertices and edges? If you look again at your coloured lines you should be able to see that they form this very shape.

Now repeat the whole operation with another colour, joining the remaining vertices together. What shape have you found this time? This means that hidden away inside the rhombic dodecahedron are a cube and an octahedron, interlaced. You have seen a sketch of this in Fig. 22. Make one for yourself if you have not already done so. Here is a brief guide. You know enough now to fill in the details for yourself.

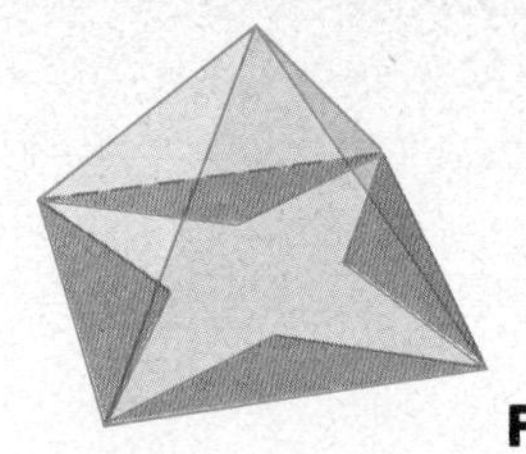

Fig. 28. Square pyramid to be stuck on one face of the cube

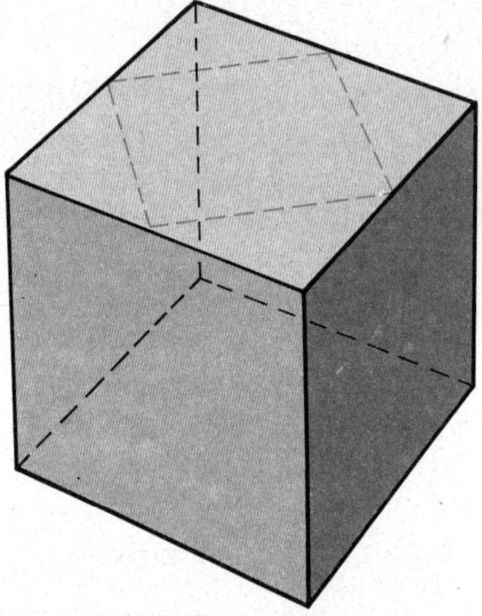

Fig. 27. Cube, showing the position of the base of the pyramid on one face

A cube is made first. Then a pyramid of four equilateral triangles is needed for each face. It should not be difficult for you now to measure the right length for the sides of the triangles.

If you stick the pyramids on to the cube they will seem to form an octahedron partly inside the cube and partly sticking out.

Write down the name of the shape where the two solids interlace.

Emphasize this shape by sticking coloured sticky tape along the creases.

Now join the vertices of your model using balsa wood strips or bright string or wire. What shape is formed? Had you expected a rhombic dodecahedron?

This latest model, shown complete in Fig. 29, is really four solids in one. Its core is a cuboctahedron (shown in outline). Round this are the cube and octahedron (in card) and enclosing them all is the rhombic dodecahedron (in balsa wood). This all happens because the semi-regular cuboctahedron and the semi-regular rhombic dodecahedron are DUALS. If you have forgotten what this means look at page 15 again.

Fig. 29. Cube and octahedron interlaced. The strips joining the vertices form a rhombic dodecahedron

This is not the first time we have come across two interlacing solids with a hidden core and its dual round the outside. Do you remember the stella octangula (page 20)? Find the models you made of this solid.

It is made of two tetrahedra and where they meet is an octahedron. If you join up the corners of the stella octangula you will find that you have a cube which is the dual of the inner octahedron.

If you made the skeleton model from Fig. 16 you will find that you have made a model of this already. If not, make one now, working it all out for yourself. You could use bright card for the two tetrahedrons, sticky tape for the outline of the inner shape and balsa wood for the cube. A model made like this is shown in Fig. 30.

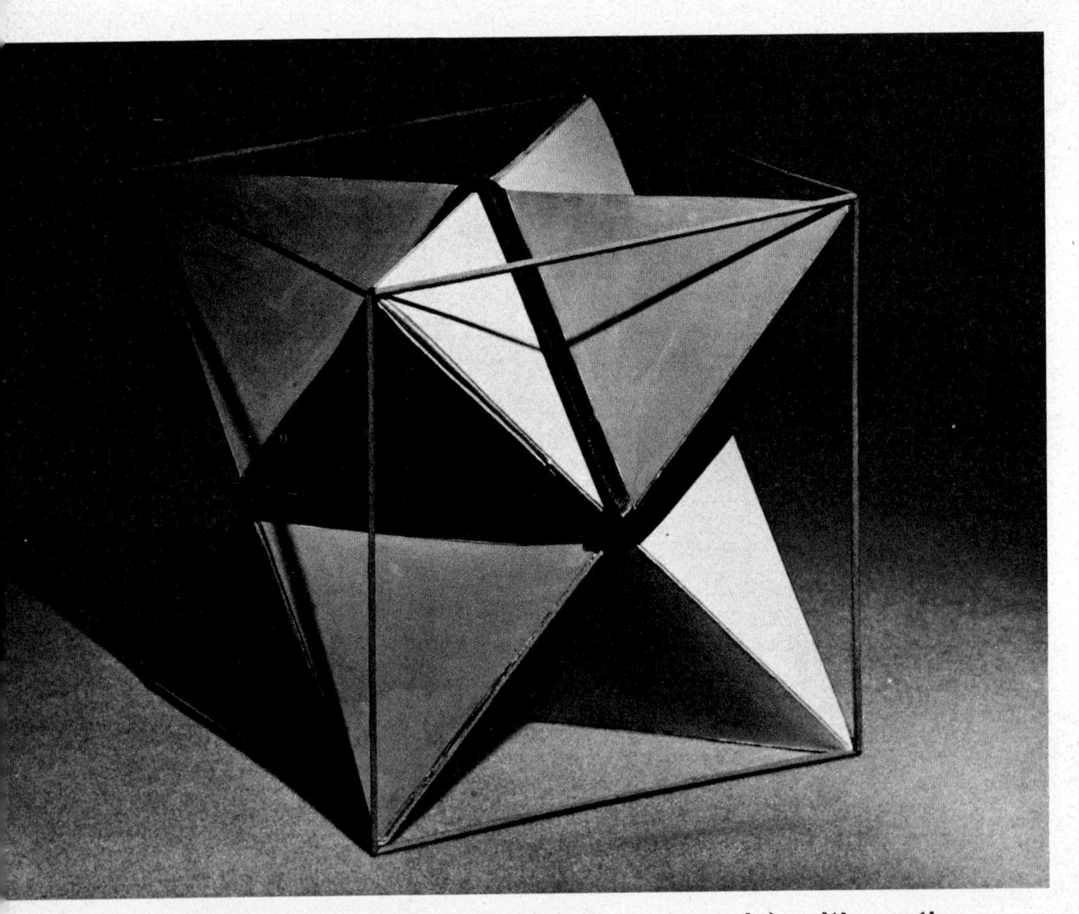

Fig. 30. Two tetrahedra (stella octangula) with vertices joined to form a cube